ABSTRACT

The first section is a brief review of the main ideas of relativistic cosmology and steady-state cosmology. The second section is a survey of observational work in optical and radio astronomy that bears directly upon cosmological problems. It is noted that much of the observational material seems to contribute to a simple coherent picture of the cosmos. But there remain serious gaps in this picture, and there are some observational results that cannot be made to fit at present. The usual development of relativistic cosmology is then sketched and an attempt is made to assess what it can be said to have achieved. Stress is laid upon the fact that certain inferences are found not to depend much upon postulated 'initial' conditions, a circumstance that may enhance their validity. It is suggested that the theory may give significant knowledge about a large tract of the universe, although it ought not to be pressed too far e.g. to the point of inferring the occurrence of an origin in a unique 'big bang'. Indications are that the behaviour of existing matter may be treated by known physics, but that some new physics is required to deal with all problems concerning origins of matter and of galaxies.

SIGMA SERIES OF SCIENCE SURVEYS

Consulting Editor:

Sir Bernard Lovell, CBE, LLD, DSc, FRS
Professor of Radio Astronomy, Manchester University
Director, Nuffield Radio Astronomy Laboratories, Jodrell Bank

COSMOLOGY

by

Professor W. H. McCrea, PhD, ScD, DSc, FRS
Research Professor of Theoretical Astronomy
Astronomy Centre, University of Sussex

FRANCIS HODGSON

SCIENTIFIC PUBLISHERS SINCE 1884

Cosmology

First edition published December 1969 by
Francis Hodgson Limited, PO Box 74
Guernsey, British Isles

Standard Book Number 85280 012 6
Library of Congress Catalogue Card No. 79–102880

ACKNOWLEDGEMENT

The publishers of the SIGMA series of science surveys wish to acknowledge the exceptional help given in the planning stages of the series by Arthur Haslett, the first Advisory Editor, who died while on holiday in June 1969. Mr Haslett worked with enthusiasm and it is largely due to his efforts that a high standard of editorial content has been achieved. The utmost will be done to ensure that this high standard is maintained.

Made and printed in Great Britain by offset by
William Clowes and Sons, Limited, London and Beccles

COSMOLOGY

Until recently, cosmology was said to be the study of the astronomical universe in the large; in practice this meant the universe of galaxies when the galaxies were imagined as smeared out to form some sort of continuous medium. However, cosmology has lately developed in numerous different directions. There are sophisticated studies of, for instance: space-time in the large; gravitation; Mach's principle; the arrow of time; intergalactic matter; background radiation; nucleo-synthesis; statistics of quasi-stellar objects, and so forth. Of course there is a developing overall picture of the cosmos and that is what a review like this ought to be about. But there are obvious difficulties: it is hard to describe the contributions of the various branches to the general picture without presupposing some knowledge of the branches themselves; also the various contributions do not cohere very well.

In the first two sections I try to supply the general theoretical and observational backgrounds, without making an unduly clear-cut distinction between them. In the third section, I try to be more specific about relativistic cosmology and what it may claim to achieve regarding the behaviour of matter

in the large. Next I briefly sketch some current thinking on the origins of matter and the universe. As then stated in the final section, the outcome appears to be that we may hope to understand the behaviour of cosmic matter without requiring wholly new concepts, but that new concepts will be needed if we are to make progress in solving any basic problem concerned with the origin of the matter.

Theoretical background

The modern study of cosmology began with a famous paper [1] by Einstein (1917). He wanted to see if general relativity agrees with Mach's principle which asserts that the inertial properties of matter on a small scale are determined by the behaviour of matter on a cosmic scale. As one way of attacking this problem, Einstein constructed the simplest model universe admitted by the theory. Provided he retained the so-called cosmical terms in his field-relations, he could produce a model that was homogeneous and isotropic in space and static in time. Interesting properties were: (a) The 'cosmical constant' Λ in the cosmical terms is related in a simple way to the mean density of matter in the universe ρ. Thus the relativistic equations that describe the local behaviour of matter depend, through Λ, upon the large-scale distribution of matter. So in a general way these equations conform to Mach's principle. (b) Space in the model being finite but unbounded, there is no problem of boundary conditions, such as in classical physics. The model being static, there is no problem of initial conditions. (c) The model is unique in satisfying Einstein's seemingly inevitable requirements. Thus the theory was unprecedentedly successful in predicting one and only one possible universe.

Einstein's satisfaction in this situation was short-lived. de Sitter (1917) soon produced another model apparently satisfying the same general conditions.[2] It, too, depended upon a non-zero Λ, but it was empty of matter. Thus it seemed to demolish two attractions of Einstein's model, uniqueness and conformity to to Mach's principle. Actually the position was not so disastrous because de Sitter's model is not naturally interpreted as static; so it does not properly satisfy Einstein's requirements. Further, when the requirements were relaxed

so as to admit non-static models, Einstein's and de Sitter's appeared as limiting states of a single model.

Non-static models were studied first by Friedman (1922), then by Lemaître (1927), the significance of their work being first apprehended [3] by Eddington (1929). Robertson (1929) and Walker (1935) supplied a stricter mathematical formulation.[4] The work produced the infinite set of relativistic (Friedman) models. Einstein welcomed these discoveries and, for his part, proposed henceforth to discard the Λ-terms.

The most important result was the impossibility of a static Friedman universe. (Einstein's static model which exists if and only if $\Lambda > 0$ was no longer acceptable because it was found to be unstable). The theory thus required the universe to be expanding or contracting. Then Hubble (1929) announced his empirical discovery [5] of the 'expansion of the universe' (which had been to some extent foreshadowed by earlier workers). This was hailed as not only an epochal discovery about the astronomical universe but also an overwhelming vindication of general relativity that had predicted it.

Even today, the immediate aim of relativistic cosmology is to isolate a particular Friedman universe that is in satisfactory agreement with observation – or to demonstrate that this is impossible. Almost the only important extension of the early work in relativistic cosmology was that of Gamow (1948) who first showed that, if the universe behaves according to the theory, there must have been a stage in its history when radiation predominated over matter.[6]

There are deep reasons why general relativity could give a conceptually satisfactory model universe, when classical physics had failed to do so. In general relativity, matter and its spatial and temporal relationships are aspects of a single entity; this we call space-time. If we can have a space-time such that, at what we call the same (cosmic) instant two parts of space are alike then their material contents are also alike. Also the theory can deal with finite or infinite amounts of matter and there is apparently no trouble about boundary conditions. In classical physics, on the other hand, space is given independently of the matter in it, and space is infinite. Classical mechanics, without

modification, deals only with finite quantities of matter. So classical theory is unequipped to deal with a universe of matter that occupies all the space of that theory. This shortcoming was common knowledge; so the ability of general relativity to produce any self-consistent model at all was recognised as an immense advance. Yet the contrast is less than first appeared. Neither theory admits a physically significant static universe. General relativity does admit non-static model universes; classical theory has been found (Milne and McCrea 1934) to admit closely similar models with only relatively minor trouble over a boundary [(7)] (McCrea 1955). General relativity disposes of such trouble and it is generally more satisfying. Nevertheless, the classical treatment (newtonian cosmology) has helped considerably in gaining physical insight.

Most relativistic cosmological models start in a singular state, implying that it is meaningless to ask what the universe was doing before that. The emergence from that state – the so-called 'big bang' – may be called the creation of the universe, and the proper-time of a 'fundamental observer' measured from that state may be called the age of the universe in the vicinity of the observer. Some thinkers regard the situation as natural and acceptable. Others find it unsatisfactory and seek some alternative. Bondi and Gold (1948) and Hoyle (1948) proposed that we should contemplate the possibility of the universe in the large being in a steady state.[(8)] This would require the continual creation of new matter throughout the universe to compensate the dispersal, by the expansion of the universe, of existing matter. Thus the hypothesis of the one cosmic event of the creation of the universe was to be replaced by the hypothesis of as many as possible uncaused events of the creation of particles of matter through all space-time. The age of the universe was to be replaced by the mean age of the contents of any part of the universe.

The steady-state hypothesis gave an effectively unique cosmological model, which was extremely vulnerable to observational test. As a matter of history, until about 1966 no observational test had altogether indubitably contradicted the model. However, by then the cumulative weight of the evidence seemed to be overwhelmingly against it.

Observational background

Optical astronomy When Hubble first published his 'law' that the redshift is proportional to the distance of the 'spiral nebulae' (galaxies) this was almost all that was known about the universe in the large. If a spectral line can be identified, its redshift z can be measured with high accuracy. The measurement of the distance of a remote object is, however, subject to large uncertainty. Thus, if according to Hubble's law the apparent speed of recession of a galaxy is H times its distance, in 1929 Hubble obtained $H \approx 550$ km/s per megaparsec (Mpc), while Sandage's (1968) latest re-determination [9] gave $H \approx 75{\cdot}3$ km/s Mpc ($H^{-1} \approx 13 \times 10^9$ years) although he says it could be as small as $H \approx 50$ km/s Mpc ($H^{-1} \approx 19 \times 10^9$ y).

In modern cosmology it has usually been assumed that there is no preferred position in the universe, the assumption in the precise form being known as the cosmological principle. Again it seems to have been Hubble (1936) who inferred this as an empirical property of the actual universe; he called it the 'principle of uniformity'. Further, if we see objects T light years away, we see them as they were T years ago; if the universe is evolving, a view of distant parts is therefore a view of an earlier state. So counts of galaxies in various ranges of apparent magnitude should help us to infer a particular cosmological model to fit the actual universe. Hubble, and others, sought also to exploit this possibility.

By 1948 no one had succeeded in fitting a plausible model to the observations. The most immediate reason was the mistaken value of H that was bound to lead to absurdly small estimates of the age of the universe. Indeed, the age difficulty originally prompted the idea of steady-state cosmology. Had Baade's (1952) first drastic revision of the value of H come before the steady-state idea had been propounded, the idea might not have gained much hearing. There would still have been an age difficulty, but it would not have been so great as to encourage people to entertain such a revolutionary remedy.

At the same time it came also to be appreciated that counts of galaxies did not give reliable statistics for the purpose in hand. Amongst other reasons, the clustering of galaxies greatly complicated the exercise. If automated techniques can be applied, it may someday be tackled afresh; but at present not even a start has been made.

So far as exclusively optical observations are concerned, efforts have consequently been largely concentrated upon obtaining the most accurate, extensive and significant empirical ('Hubble') relation between log z and m, where m is the apparent magnitude, on some standard scale, of a galaxy exhibiting redshift z. We require the relation for some standard class of galaxy; it is widely considered that we do best to use the brightest galaxy in a cluster. If the universe is evolving (not in a steady state) then the more remote objects observed are in general younger than nearby objects; so objects that are otherwise standard must in general show an age-distance effect. Since, however, for purely 'optical' galaxies the greatest values of z are about 0·2, the corresponding objects are less than about 3×10^9 light years away. We expect the luminosity of most large galaxies not to have changed considerably in about the past 3×10^9 years. So here we expect no considerable age-distance effect. For objects exhibiting larger redshifts (see below) the effect may be important.

Radio astronomy The greatest development in observational cosmology during the past 20 years has been the application of radio astronomy and the work in optical astronomy arising therefrom. Radio astronomers have shown that a considerable proportion of all observed discrete radio sources are 'radio galaxies' and optical astronomers have identified many of these with optical objects. These are evidently galaxies in some relatively unusual state which results in their producing radio emission almost a million times as intense as that from normal galaxies like the Milky Way. Optically they are about as luminous as other bright galaxies, but emission lines are a more notable feature in their spectra.

Such studies led to the conclusion that radio astronomy could explore the universe to much greater depths than optical astronomy. Also it yielded apparently significant counts (N) of sources down to various flux-levels (or radio apparent magnitudes) of the sort that had defeated optical astronomy. Various counts made at Cambridge are the best known[10] (Ryle 1968). These appear by large factors to contradict the predictions of steady-state cosmology. Their interpretation according to evolutionary cosmology leads to the inference that radio sources were much more abundant in a cosmic era corresponding to $z \approx 3$ to

4 than they are at present, while still earlier at $z \approx 4$ to 8 they were much less abundant (these z-values being extrapolated from nearby sources) — so that radio astronomy may already have penetrated to the limit of the radio universe. Thus the counts are apparently so much affected by the evolutionary history of the sources that they are not of positive help in selecting a particular cosmological model. Another handicap is that radio astronomy has not yet produced any way of measuring a redshift — we should naturally have greater confidence in the interpretation of the counts if the redshifts of the faint sources could be measured directly.

Other counts made at shorter wavelengths have given seemingly discordant results and it has been claimed that they do not conflict with steady state cosmology[(10)] (Shimmins, Bolton and Wall 1968). However the Cambridge observers believe that the new results actually confirm their own (Pooley 1968).

QSOs A quasi-stellar object (QSO) is an object that possesses most of the several features that collectively distinguish QSOs from all other objects in the heavens. Some (quasars) are radio sources, generally of exceedingly small angular size. Of the few hundred apparently strongest radio sources, roughly one-half are quasars; of the fainter sources in the Cambridge counts, probably a majority are quasars. Most QSOs show strong emission lines in the optical spectrum. Many of these show redshifts far exceeding any found for other objects; the largest so far is $z \approx 2{\cdot}4$, which means that the lines are seen at 3·4 times their standard wavelengths! If these large redshifts are 'cosmological', in the sense of having the same significance as the Hubble redshifts, then these QSOs are by far the most remote objects known. Also they must be by far the most luminous. In that case they enable astronomers to explore about 1000 times the volume of the universe otherwise observable. This is the main cosmological interest of QSOs.

The surprising feature (Burbridge and Hoyle 1966, 1967) is that the apparent magnitude (optical or radio) is not correlated with the redshift as it is in the Hubble relation for galaxies at smaller redshifts.[(11)] Maybe intrinsic scatter or evolutionary effects mask the expected correlation. Or maybe the universe is more complex beyond the range of the classical Hubble observations.

But some astronomers infer the existence of some cause of redshift besides the expansion of the universe. Further evidence for this seems to come from cases[12] of two or more objects that are apparently physically related but have widely different redshifts. (Arp 1967, Sargent 1968). On the other hand, some radio astronomers find that, as regards every characteristic property, radio galaxies and quasars show some overlap, if and only if quasars are at 'cosmological' distances. This is the so-called 'continuity argument' in favour of the cosmological interpretation of z.

Intergalactic matter The universe is presumably made of matter and electromagnetic radiation. In the observable universe, the mass of the radiation is small compared with that of the matter, and intergalactic space is largely transparent to the radiation.

All the matter detected by existing astronomical instruments is matter of galaxies (and QSOs). But astronomers have long appreciated that there could be much more matter spread through intergalactic space in forms that would evade detection by means hitherto employed. It might be diffuse gas of neutral or ionized atoms, cosmic rays, solid material in various forms, or even an all-pervading ocean of neutrinos. Matter of some such sort could well amount to 1000 times the amount of matter in galaxies without having yet been detected. So we ask, is the astronomical universe composed of the rather untidy distribution of galaxies and almost nothing else? Or does it consist of a more even distribution of a greater amount of matter that we do not yet know about, in which the matter we do know about in the form of galaxies amounts to only minor random irregularities?

These are the most pressing questions for observational cosmology; Gould (1968) has recently reviewed this now complex subject.[13] The current estimate of the contribution of matter in galaxies to the mean density is 7×10^{-31}g cm^{-3}. The simplest Friedman universe gives a total mean density about 2×10^{-29}g cm^{-3}; this is about the largest estimate that is taken seriously. If such an excess of non-galactic matter exists, it is thought most likely to be mainly ionized hydrogen. But no significant amount of non-galactic matter has yet been detected beyond doubt.

Background radiation Radio observations at centimetre wavelengths give substantial support to the inferred existence of a cosmic background radiation that, at our epoch in our cosmic neighbourhood, is black-body radiation close to 2.7°K (Penzias and Wilson 1965, 1968 and many others).[14] The universe is highly transparent at the wavelengths concerned, so the radiation must have travelled much further than the distance of the remotest objects we see, in order to reach us now, and must have suffered the corresponding redshift. Thus, when the background radiation was produced, it was of much shorter wavelength and had a much higher temperature than it has now, i.e. the universe was in a very different state. Accordingly the existence of this radiation is perhaps the strongest observational evidence for an evolutionary, and not a steady-state, universe. Carrying the argument further leads to the inference that the universe started from a 'hot big bang' or 'universal fireball'.

Further, if the radiation is indeed black-body, it serves to establish as a preferred frame of reference at any event that in which the radiation is isotropic. Observers claim that the observed radiation-flux is isotropic to within departures of less than 0·2 percent; this may have interesting implications. However, the chief interest of the radiation is simply that it adds a new fundamental item to our quantitative knowledge of the cosmos.

Summary To summarize the empirical position, we find one category of observations that apparently fit together to give a simple coherent picture; but we find other observations, and some lack of observations, that suggest we may still be a long way from any such picture. The first category includes the Hubble relation for optical galaxies, most of the radio source counts, the background radiation if the black-body character be accepted, and the various instances of continuity of properties of galaxies, radio-galaxies and QSOs. The contrasting category includes the apparent lack of a Hubble relation for QSOs, counts that may disagree with previous radio-source counts (or the lack of information about the spectra of the sources needed to say whether they do disagree), the lack of direct observations of the background radiation in the most critical wavelength interval, and the lack of tests of its isotropy on a

small scale (say ~ 1 arc minute) or on a large scale (say between the northern and southern hemispheres), the lack of decisive observations of intergalactic matter.

Relativistic cosmology

Relativistic cosmology (like most cosmological theories) seeks to construct a theoretical model of a smoothed-out universe, i.e. the aim is to reproduce the broad features of the actual universe got by averaging observed masses over appropriate volumes, and so on. The theory usually requires the model to satisfy postulates of homogeneity – the cosmological principle – and isotropy. In the first instance, these postulates may be no more than specifications of the simplest possible cases for investigation.

It then follows that the model can be treated as its own clock reading 'cosmic time' t, and it is found to be specified completely by an 'expansion factor' $R(t)$ and a constant k having one of the values $k = +1$ (closed model) or $k = 0$ or -1 (open models). The density $\rho(t)$ and pressure $p(t)$ of the contents of the model are then expressible in terms of $R(t)$ and its first and second derivatives, and the constants k and Λ (cosmological constant). Any postulated equation of state of the contents, in the form of a relation between ρ and p, then yields a second-order differential equation for $R(t)$. The relation normally used is simply $p = 0$, it being noted that the universe as observed is dominated by matter having negligible mean pressure.

The solution of the differential equation introduces one significant arbitrary constant (and a constant for the origin of t). Treating Λ as a further disposable constant, we obtain a 2-parameter family of (Friedman) models. In comparing theory and observation we aim to evaluate the two parameters that determine a model, together with the epoch $t = t_0$ at which the observations are made.

We use suffix zero for variables evaluated at $t = t_0$. The most convenient parameters to determine the model are the dimensionless numbers

$$\sigma_0 = \frac{4\pi G \rho_0}{3H_0^2} \qquad q_0 = -\frac{\ddot{R}_0}{R_0 H_0^2} \qquad (G = \text{gravitation constant})$$

where

$$H_0 = \dot{R}_0/R_0 \quad \text{is Hubble's constant for the observer.}^{(15)}$$

We think of q_0, σ_0 as coordinates in a plane; any point (q_0, σ_0) denotes a particular Friedman model, apart from a scale factor. If a value of H_0 is specified, it determines this factor and thence the value of t_0.

For any Friedman model there is a theoretical m, log z-relation depending upon H_0, σ_0, q_0 and the intrinsic properties of the objects concerned. Comparison with the empirical Hubble relation enables us to estimate H_0 and to say, with some degree of probability, that the actual universe behaves like a Friedman universe having (q_0, σ_0) lying within some particular region in the q_0, σ_0-diagram. Most other empirical properties of the cosmos impose some restrictions upon the admissible values of q_0, σ_0.

Much of the current effort in analysing the large-scale observations of the universe is to find whether any Friedman model is compatible with all the observations; more precisely, it is to find if any region of the q_0, σ_0-plane remains admissible after all empirical restrictions have been applied. It is an anticlimax to record now that the outcome of tremendous exertions is still inconclusive, because at present all the empirical information is too imprecise. Available observations do not exclude the possibility of finding an acceptable Friedman model; on the other hand, they cannot be claimed markedly to favour even any general type of model, say 'open' or 'closed' or having zero or non-zero Λ, and so on.

Assessment When we remember the elaborate mathematical apparatus needed to treat the simplest elementary particle in the universe, or the complexity of the modern study of gravitation, and when we consider all the scope there might be for new concepts in attempting to treat the universe in the large, then it seems incredible that we should hope to infer the whole history of the cosmos, from the grand event of creation on to eternity, by using one single ordinary differential equation. Also the form of the equation employed depends upon the seemingly incidental feature of the smallness of the mean pressure of the contents of the universe. Yet, taken literally, this is what the theory outlined above is attempting.

The following is perhaps a fair assessment: Because it is the only course available, cosmologists take a grossly simplified treatment of the universe and hope it will show in broad terms how some large part of the universe behaves through some large interval of its history. The only characteristic time in the subject is the Hubble time H_0^{-1} and so a large time should be of this order and a large region should have linear extent of order cH_0^{-1}. Anything we validly assert about the 'universe' might be expected to apply to a region up to about this extent; this is apparently what some writers call the 'metagalaxy'.

According to the theory the universe cannot be static. According to observation and the theory, it is expanding from some highly condensed state, this expansion proceeding under the gravitational interaction of the material alone. According to observation, the universe is matter-dominated at present. According to the theory, there must have been an earlier stage when it was radiation-dominated. According to observation and the theory, some of the radiation survives in the form of the present background radiation. According to the theory, there was an early stage when the density and temperature were so high that the time-scale of nuclear reactions was small compared with the time-scale of the expansion; there was then thermal equilibrium among nuclear species. Therefore the relative abundances of nuclear species at any subsequent epoch are independent of nuclear features before that stage, i.e. there would be no meaning in postulating any 'initial chemical composition'. The abundances of some light nuclei are largely those that became frozen in following that stage. Such abundances depend upon the temperature and density at that stage and these are related to the present values. In particular, it follows that the cosmic abundance of helium, the temperature of the background radiation, and the mean density of the universe should satisfy a certain relation (depending upon σ_0, q_0). Wagoner *et al.* (1967) find plausible results from extensive computations on this subject.[16] This account also leaves the required scope for nucleosynthesis of heavier nuclei in stars as the universe evolves.

Some authors have discussed the behaviour of the universe even earlier than the stage mentioned above. One inference is that there was a high neutrino density when the temperature was over 10^{10} °K. Misner (1968) made the remarkable

discovery that, were the universe then highly anisotropic, 'neutrino viscosity' would rapidly induce a high degree of isotropy.[17] The theory still offers no explanation of the start of the expansion, but Misner's work tends to show that what we observe does not depend much upon the way it started.

In ordinary physical problems we seek a solution subject to given boundary conditions. In the cosmological problem boundary conditions cannot be given; if they have a meaning, it is part of the problem to find them. It has often seemed that the only hope of progress using normal physical procedures would be to discover that the solution is almost independent of any particular boundary conditions assumed. Such a state of affairs has not yet been realised. But perhaps the most significant aspect of recent work is that it points this way: the predicted chemical composition does not depend upon any particular initial composition; the predicted isotropy does not depend upon any particular initial isotropy.

Cosmogony

Any Friedman model that serves for the foregoing discussion is one that starts with a 'big-bang'. In the model this is a unique event that involves the whole universe. While a model of this sort, rather surprisingly, yields a coherent account of many features of the cosmos, it seems to do so precisely because these features do not depend too much upon the assumed starting conditions. Therefore we must not assign too much significance to such conditions, particularly to the big bang. Also there are specific objections: most big-bang models give an implausibly small 'age of the universe'; there is evidence that there does not exist a cosmic helium abundance compatible with a single big bang; and so on. Indeed, many of the reasons that led to the consideration of continual creation and steady-state cosmology seem to persist, although observational objections to the steady-state model seem insuperable.

Big-bang cosmology has the utmost uniformity in space and the utmost non-uniformity in time — the unique uncaused big-bang for the creation of the whole universe. Steady-state cosmology went to another extreme in requiring the

utmost uniformity in time as well as in space – with an uncaused 'micro-bang' for the creation of every elementary particle. There remains the possibility of some non-uniformity both in space and in time – any number of lesser-bangs occurring at random throughout space and time.

The work of several authors contains some element of this concept. Jeans (1928) speculated about the central parts of galaxies being places where new matter is injected into the universe;[18] Zelmanov (1960) has studied space-times incorporating both expanding and contracting regions – maybe he could admit the possibility of an expanding region starting from its own initial 'bang'; McCrea (1964) considered the possibility that matter might not be strictly conserved in in certain reactions so that we might have to consider a self-propagating universe;[19] Hoyle and Narlikar (1966) made a more precise proposal about non-conservation of baryon number;[20] more recently, Hoyle (1969) has revived something like Jeans' idea;[21] Harrison (1968) has discussed the role of primeval irregularities in connection with the formation of condensations and with the occurrence of matter and antimatter.[22] However, the astronomer who has most consistently advanced ideas along these lines[23] and supported them by empirical evidence is Ambartsumian (1961). He calls attention particularly to groups of galaxies that appear to have positive total energy. They would then be dispersing ('expanding') systems presenting an acute problem as to how they started. Whereas most hypotheses about the origin of galaxies regard them as being formed by the condensation of more tenuous material, Ambartsumian also remarks that in practice astronomers never observe instances of this, but they do see plenty of occurrences of the opposite sort – of matter emerging from condensations. Ambartsumian refrains from speculating about the origin of the matter in such cases. But, if appearances are not deceptive, they indicate the operation of some process not otherwise known in physics; its inception would be the sort of 'lesser bang' we have contemplated.

Such ideas are not yet incorporated into a proper theory. Perhaps the main empirical evidence supporting them is the apparent lack of correlation of redshift and apparent magnitude for quasars[11] and the sort of evidence[12] collected by Arp. The main empirical evidence against the ideas would be the

uniformity and isotropy of the universe in the large, if the evidence is valid. Actually these are features that should be critically re-examined. For instance, the existence of a feature like the de Vaucoulers super-cluster might suggest large-scale inhomogeneity.[24] Again, Faulkner, *et al* (1966) find a remarkable apparent anisotropy in the directions in which we see quasars of high z-values.[25]

The ideas in this section are unfortunately vague. So far as they go, they have the merit of combining the broad advantages of evolutionary and of steady state cosmology. Also they are almost certainly bound up with two further basic problems of cosmogony:–

a) *Matter and antimatter* Since matter and anti-matter have symmetric properties, it is natural to expect the universe to contain comparable amounts of the two sorts. It is impossible to see how this could be achieved by one simple big bang. If we accept that there has been such a bang, we should wish to know why it produced one sort of matter and not the other. On the other hand, if we have a number of lesser bangs, some might produce one sort and some the other, thus yielding symmetry in amounts or in probability of production. Another deep problem[26] in the same general category is, Why are all electrons alike, all protons alike, . . . ? (Clerk Maxwell 1873).

b) *Galaxy formation* Ideas regarding the formation of galaxies have ranged from condensation out of a uniform distribution of diffuse matter in an evolving universe, through fresh condensations induced by previously existing condensations in a steady-state universe, to something like special creation in a multi-bang universe. The problem is difficult to formulate because a uniform distribution of raw material is the most improbable state imaginable, while if we admit non-uniformity at whatever stage we start from, we necessarily somewhat beg the question as to the formation of condensations! There seems to be no avoiding the inference that irregularities of density must have existed at all epochs, so that the problem of the origin of galaxies is necessarily bound up with the problem of the origin of the universe itself. The question is the extent to which this leaves anything to be discussed apart from this overriding problem. Peebles (1968), for instance, seems to have made interesting progress using a big-bang model. However, it is still not clear whether such an approach tells us,

say, why the galaxies have the masses they do have.

Conclusion

The separation of the ideas of the last two sections seems to be fundamental. The substance of the first of them, dealing with the *behaviour* of matter in the large, has the makings of a considerable body of knowledge. Its full working out may require some elaboration of existing physics, but there is no indication that any new physics is needed for the problems concerned. The work of many investigators in this field may have produced smaller progress than could have been hoped, but it has never produced any absurd result, which it surely would do were it fundamentally mistaken or deficient. The second of these sections, dealing with questions of *origins*, is a different story. While we may be able to say something about the sort of origins to look for, we still have no theory of them.

In passing from matter of everyday experience to micro-matter, we meet in quantum physics a new set of concepts concerning the behaviour of matter. In passing from everyday matter to cosmic matter, apparently we do not meet new concepts about its behaviour. Where we must apparently expect new concepts – and we have little idea as to what they may be like – is in the problem of origins.

If this conclusion is valid, it is an advance in insight. Briefly it implies:

> Relativistic cosmology, together with existing physical theory, gives a meaningful account of the behaviour of some large part of the cosmos as regards motion, element-production, proportion of matter and radiation, and so on. New concepts will be demanded to deal with all problems of origins from those of elementary particles to those of any other quantities of matter, present indications being that the 'creation' of new matter may be a more fitful process than hitherto envisaged.

Finally, we recall that Einstein's interest in Mach's principle sparked off all this work. A related topic is that of the 'arrow of time' – attempts have been made to show that the asymmetry between past and future is related to the one-way expansion of the universe. The interest of such topics is the possibility

of a connection between the behaviour of bits of matter that we can handle and the behaviour of the universe on the grand scale. The formulation of the problems has been clarified but the problems have not yet been solved.(27)

REFERENCES

1 Einstein, A. Cosmological considerations on the general theory of relativity in *The Principle of Relativity*, A. Einstein *et al.* (Methuen, London 1924; also Dover, New York) pp 177–188

2 de Sitter, W. On Einstein's theory of gravitation and its astronomical consequences. *Monthly Notices of the Royal Astronomical Society*, **78** 3–28 (1917)

3 Eddington, A.S. The instability of Einstein's spherical world. *Monthly Notices of the Royal Astronomical Society*, **90** 668–678 (1930)

4 Robertson, H.P., Noonan, T.W. *Relativity and Cosmology* (Saunders, Philadelphia, London, Toronto 1968)

5 Hubble, E.P. *Realm of the Nebulae* (Oxford: University Press 1936) *The Observational Approach to Cosmology* (Oxford: University Press 1937)

6 Gamow, G. *The Creation of the Universe* (Viking, New York 1952, revised 1961)

7 McCrea, W.H. On the significance of newtonian cosmology. *Astronomical Journal*, **60** 271–274 (1955)

8 Bondi, H. *Cosmology* (Cambridge: University Press, 2nd ed. 1960)

9 Sandage, Allan. A new determination of the Hubble constant. *Astrophysical Journal*, **152** L149–L154 (1968)

10 Ryle, M. Counts of radio sources. *Annual review of Astronomy and Astrophysics USA* **6** 249–266 (1968); Shimmins, A.J., Bolton, J.G., Wall, J.V. Counts of radio sources at 2,700 MHz. *Nature* London **217** 818–820 (1968); Pooley, G.G. Counts of radio sources at 2,700 MHz. *Nature* London **218** 153–154 (1968)

11 Burbidge, G., Burbidge M. *Quasi-stellar Objects* pp 154, 161 (Freeman, San Francisco and London 1967)

12 Arp, H. Peculiar galaxies and radio sources. *Astrophysical Journal* **148** 321–365 (1967); Sargent, W.L.W. Redshifts of galaxies in the remarkable chain VV 172. *Astrophysical Journal* **153** L135–L137 (1968)

13 Gould, R.J. Intergalactic Matter. *Annual review of Astronomy and Astrophysics USA* **6** 195–214 (1968)

14 Chiu, H.–Y. The evolution of the universe. *Science Journal* **4** 33–38 (1968)

15 Stabell, R., Refsdal, S. Classification of general relativistic world models. *Monthly Notices of the Royal Astronomical Society* **132** 379–388 (1966)

16 Wagoner, R.V., Fowler, W.A., Hoyle, F. On the synthesis of elements at very high temperatures. *Astrophysical Journal* **148** 3–49 (1967)

17 Misner, C.W. The isotropy of the universe. *Astrophysical Journal* **151** 431–457 (1968)

18 Jeans, J.H. *Astronomy and Cosmogony* p352 (Cambridge: University Press 1928)

19 McCrea, W.H. Continual creation. *Monthly Notices, Royal Astronomical Society,* **128** 335–344 (1964)

20 Hoyle, F., Narlikar, J.V. On the effects of the non-conservation of baryons in cosmology, etc. *Proceedings Royal Society,* London **A 290** 143–161, 162–176, 177–185 (1966)

21 Hoyle, F. Highly condensed objects. *Quarterly Journal, Royal Astronomical Society,* **10** 10–20 (1969)

22 Harrison, E.R. The early universe. *Physics Today,* **21** 31–39 (1968); Novikov, I.D., Zeldovic Ya. B. Cosmology. *Annual review of Astronomy and Astrophysics,* **5** 627–648 (1967)

23 Ambartsumian, V.A. Problems of extra-galactic research. *Transactions of the International Astronomical Union* XI B 145–160 (1961)

24 Weigert, A., Zimmermann, H. *ABC of astronomy,* p122. (London: Hilger and Watts 1967)

25 Strittmatter, P., Faulkner, J., Walmsley, M. Evidence for a correlation between the position and redshift of quasars. *Nature* London, **212** 1441–2 (1966)

26 Maxwell, J.C. *Scientific Papers* vol 2. p375 (Cambridge: University Press 1890)

27 Hogarth, J.E. Cosmological considerations of the absorber theory of radiation. *Proceedings Royal Society London* **A 267** 365–385 (1962); Lynden-Bell, D. On the origins of space-time and inertia. *Monthly Notices of the Royal Astronomical Society,* **135** 413–428 (1967); Sciama, D.W. *On the unity of the universe.* (London: Faber and Faber 1959)

William Hunter McCrea FRS, Research Professor of Theoretical Astronomy in the University of Sussex since 1965. Born in Dublin and educated at Chesterfield School, Trinity College, Cambridge, and Göttingen University. Has been a Bye-Fellow of Gonville and Caius College, Cambridge and is a Fellow of the Imperial College of Science and Technology in London. Has held appointments in departments of mathematics in the Universities of Edinburgh, Belfast and London (Imperial College and Royal Holloway College). Was Temporary Principal Scientific Officer, Admiralty 1943–5 and was commissioned in RAF, VR (Training Branch) 1941–45. Has been visiting professor in various departments of astronomy in USA and visiting lecturer in several departments of astronomy in Europe. President of the Royal Astronomical Society 1961–63, President of Section A, British Association 1966. Main present astronomical interests are in evolution of stellar systems and in cosmology.

SIGMA SCIENCE SURVEYS

Consulting Editor: Professor Sir Bernard Lovell, OBE, LLD, DSc, FRS

Physics in Archaeology - M. J. Aitken, MA, DPhil, Research Laboratory for Archaeology, Oxford

Towards Fusion Reactors - Academician L.,A. Artsimovich Academy of Sciences of the USSR, Moscow

Status of General Relativity - Professor W. B. Bonnor, PhD, DSc, Queen Elizabeth College, University of London, London

Rainmaking - E. G. Bowen, CDE, MSc, PhD, FAA, Chief, Division of Radiophysics, Commonwealth Scientific and Industrial Research Organisation, Sydney

Theory of the Atomic Nucleus D. M. Brink, DPhil, Department of Theoretical Physics, University of Oxford, Oxford

Superconductivity - J. A. Catterall, PhD, Division of Materials Applications, National Physical Laboratory, Middlesex

Theory of Fluids - Professor G. H. A. Cole, PhD, DSc, FRAS, Department of Physics, University of Hull, Hull

Advances in Electron Microscopy - V. E. Cosslett, MA, PhD, ScD, Cavendish Laboratory, Cambridge

Solar Wind - Professor J. W. Dungey, MA, PhD, Department of Physics, Imperial College, of Science and Technology, London

Variable Stars - Professor O. J. Eggen, PhD, Mount Stromlo and Siding Spring Observatories, Australian National University, Canberra

Radio Echo-sounding of Ice Sheets - S. Evans, PhD, Scott Polar Research Institute, Cambridge.

Automatic Analysis of Materials - M. P. Jones, Department of Mining and Mineral Technology, Royal School of Mines, London

Metal Physics - Professor P. B. Hirsch, FRS, Department of Metallurgy, University of Oxford, Oxford

Fibre Reinforced Materials - Anthony Kelly, ScD, Superintendent, Division of Inorganic and Metallic Structure, National Physical Laboratory, Middlesex

Human Engineering - Professor R. M. Kenedi, Department of Mechanical Engineering, University of Strathclyde, Glasgow

Solar Disturbances - Professor Dr K. O. Kiepenheuer, Fraunhofer Institut, Freiburg-im-Breisgau, Germany

Climatic Fluctuations - H. H. Lamb, MA, Meteorological Office, Bracknell, Berkshire

Cosmology - Professor W. H. McCrea, PhD, ScD, DSc, FRS, Astronomy Centre, University of Sussex, Falmer, Sussex

Hot Atom Chemistry - A. G. Maddock, ScD, Cambridge University, Lensfield Road, Cambridge

Numerical Weather Forecasting - B. J. Mason, DSc, FRS, Director General, Meteorological Office, Bracknell, Berkshire

The Mid-Atlantic Ridge - R. G. Mason, PhD, Geophysics Department, Imperial College of Science and Technology, London

Atomic Physics of the Upper Atmosphere - Professor Sir Harrie Massey, FRS, Department of Physics, University College, London University, London

Polymer Science - Sir Harry W. Melville, KCB, DSc, FRS. Principal, Queen Mary College University of London and former Secretary, Department of Scientific and Industrial Research, London

Dating by Radiosotopes - S. Moorbath, DPhil, Department of Geology and Mineralogy, University of Oxford, Oxford

Glacier Research - Professor J. F. Nye, MA, PhD, H. H. Wills Physics Laboratory, Bristol University, Bristol

Pattern Recognition - John R. Parks, PhD, Division of Computer Science, National Physical Laboratory, Teddington, Middlesex

Infrared Astronomy - Professor J. Ring, DSc, PhD, Department of Physics, Imperial College of Science and Technology, London

Polar Ice Sheets - G. de Q. Robin, MA, PhD, Director, Scott Polar Research Institute, Cambridge

Theory of Sampling - Professor M. R. Sampford, MA, DPhil, Department of Computational and Statistical Science, University of Liverpool, Liverpool

Pulsars - Professor F. Graham Smith, Nuffield Radio Astronomy Laboratories, Jodrell Bank, Cheshire

Field-ion Microscopy - M. J. Southon, PhD, Department of Metallurgy, University of Cambridge, Cambridge

X-ray Astronomy - Professor E. A. Stewardson, MSc, PhD, Department of Physics, University of Leicester, Leicester

Continental Drift - Professor J. Sutton, PhD, FRS, Imperial College of Science and Technology, Department of Geology, Royal School of Mines, London

Evolution of Stars - Professor R. J. Tayler, PhD, Astronomy Centre, University of Sussex, Falmer, Sussex

Photogrammetric Methods - Professor E. H. Thompson, OBE, MA, FRICS, Department of Photogrammetry and Surveying, University College, London University, London

Cosmic Radiation - Professor J. G. Wilson, MA, PhD, Department of Physics, University of Leeds, Leeds

Star Clusters - Professor Sir Richard Woolley, OBE, ScD, LLD, DPhil, FRS, Astronomer Royal, Royal Greenwich Observatory, Hailsham, Sussex

SIGMA SURVEYS

The object of the SIGMA survey series is to provide scientists and technologists with up to date knowledge of developments in fringe areas of research borderline to their main interest. Each survey reviews the present position of a selected field of science and current and recent achievements over the previous year or more and these activities are projected into the immediate future. The surveys are limited to a length of between 5,000 and 8,000 words in order to increase their chance of being actually read and are normally published within a few weeks of being written.

Status of General Relativity by Professor W. B. Bonnor, PhD, DSc, Queen Elizabeth College, University of London, London

Cosmology by Professor W. H. McCrea, PhD, ScD, DSc, FRS, Astronomy Centre, University of Sussex, Falmer, Sussex

Ready for publication early in 1970

Superconductivity by J. A. Catterall PhD, Division of Materials Applications, National Physical Laboratory, Middlesex

Theory of Fluids by Professor G. H. A. Cole, PhD, DSc, FRAS, Department of Physics, University of Hull, Hull

Solar Wind by Professor J. W. Dungey, MA, PhD, Department of Physics, Imperial College of Science and Technology, London

Variable Stars by Professor O. J. Eggen, PhD, Mount Stromlo and Siding Spring Observatories, Australian National University

Fibre Reinforced Materials by Anthony Kelly, ScD, Division of Inorganic and Metallic Structure, NPL, Middlesex

Numerical Weather Forecasting by B. J. Mason, DSc, FRS, Director General, Meteorological Office, Bracknell, Berkshire

Polymer Science by Sir Harry Melville, KCB, DSc, FRS, Queen Mary College, University of London and former Secretary, DSIR

Dating by Radiosotopes by S. Moorbath, DPhil, Department of Geology and Mineralogy, University of Oxford, Oxford

Francis Hodgson Limited

Scientific and Technical Publishers since 1884

PO BOX 74, GUERNSEY, CI, BRITISH ISLES